YOUR KNOWLEDGE HAS VALUE

- We will publish your bachelor's and master's thesis, essays and papers

- Your own eBook and book - sold worldwide in all relevant shops

- Earn money with each sale

Upload your text at www.GRIN.com and publish for free

Natural Disaster Relief Mechanisms in Canada, Bangladesh and Indonesia

Leo Kempe

G R I N

Bibliographic information published by the German National Library:

The German National Library lists this publication in the National Bibliography; detailed bibliographic data are available on the Internet at http://dnb.dnb.de.

ISBN: 9783346999801
This book is also available as an ebook.

© GRIN Publishing GmbH
Trappentreustraße 1
80339 München

All rights reserved

Print and binding: Books on Demand GmbH, Norderstedt, Germany
Printed on acid-free paper from responsible sources.

The present work has been carefully prepared. Nevertheless, authors and publishers do not incur liability for the correctness of information, notes, links and advice as well as any printing errors.

GRIN web shop: https://www.grin.com/document/1441565

University of Illinois at Urbana-Champaign   Leo Kempe

Fall Term 2022

Research Paper – Final Version

Policy to Recover – Natural Disaster Relief Mechanisms in Canada, Bangladesh, and Indonesia

1. Introduction

Climate change is among the most pressing challenges our planet and society must face. The increase in temperatures impacts staple crop yields, water accessibility, or health status but also the environment: Storms, floods, and earthquakes are increasing in both intensity and scope which demands more holistic and effective disaster relief concepts and operational mechanisms. Of course, this has implications for policies of affected countries, which often do not have the necessary material resources and kinds of capital (political, financial, social) that are needed.

This paper analyzes four individual kinds of natural hazards, circumstances that adversely affect the livelihoods of residents and the environment. These are wildfires, floods, cyclones, and earthquakes. The types account for geographical or spatial variation in the realm of disaster analysis since each of them could be more prevalent in specific areas. Moreover, with the four kinds, the paper distinguishes between the developed and the developing world: Canada is chosen as exemplary country to examine the first two kinds, while Bangladesh and Indonesia are used to explain the last two, respectively.

Examining different case studies from these countries, the paper highlights policy mechanisms and cooperation efforts each country has undertaken or aims to undertake to fight the most prevalent disasters. It asks through what means and by which actors these will be achieved to underline the motivations for disaster relief efforts, enhancing the understanding of different approaches to the issue. Building on these questions, the paper argues that resilient, strong, and responsive institutions are important in all example countries to foster adequate coping strategies when experiencing a natural disaster. Having adequate and effective frameworks in place drives necessary transnational cooperation between societal and political actors, something indispensable when trying to achieve synergistic effects. Since geospatial, social, cultural, political, and other situational characteristics are fundamentally different in the three countries selected, the paper is not intended to be a one-sided comparison. However, it emphasizes overlaps and commonalities, making the topic more applicable to the larger political and geographical contexts. This encourages the identification of mentioned decision-making synergies and cross-sectional conceptual learning processes.

In the upcoming chapters, the paper illustrates the disasters in the respective countries, using research articles, governmental documents, and reports written by organizations. These documents offer a thorough view of preparation and response mechanisms in the wake of a natural disaster, incorporating a variety of institutional arrangements to facilitate the approaches. Examples from different geographical areas help to grasp the scope and scale of connectivity in striving to achieve a safer environment for a country's residents. Safety and a robust response are a strong marker of a sustainable environment and society, which is why the paper employs the United Nations Sustainable Development Goals (UN SDGs) as an underlying explanatory concept.

In a conclusion, the paper summarizes findings and illustrations of the three selected countries. Moreover, it gives an outlook as to where future research might go and how the respective regions will develop in terms of disaster adaptation tactics. This will keep the conversation about the interactions between climate change and the three pillars of sustainability going, while also providing further incentives for considering the topics in national policies and academia.

2. Country: Canada – North American Climate Change Vulnerabilities

Canada, the second-largest country on our planet Earth, faces multiple climate-related challenges, explained in this paper. One of the most severe is the threat of wildfires – fires that burn in the numerous forest, grassland, and swamp regions of the country.

Currently, the International Association of Wildland Fire (IAWF) says that this type of disaster is substantially more costly than ever (IAWF 2015). This is because with climate change, wildfires have grown in frequency, intensity, time, and scale (Canadian Climate Institute 2020). This reality, the IAWF says, begs for a more concrete analysis of the kinds of cost burdened on individuals, communities, and governments (IAWF 2015): Direct costs are readily observable, with impacts on daily life, health, property, supply chains, and the time and resources of keeping a particular fire in check. Indirect ones, however, are those more hidden to the public eye, at least in an initial consideration. These are the supplies and equipment as well as the training fire crews need, complemented by things such as insurance or the impact on localized ecosystems. Because of the temporal nature of fire disasters, there are also costs after it (rehabilitation). Evidently, then, wildfires threaten not only people and their homes, but also the animals that have their habitat in the affected regions. Hence, SDG #15 (Life on Land) becomes relevant (UNDESA n.d.), directly translating into protection and conservation efforts. It must be ensured that the threat to ecosystems and human lives is minimized (IAWF 2015), a call invoking the concept of resilience.

Typically, resilience refers to a state of organization in which humans are equipped with physical and psychological tools to face disasters like wildfires – for instance building material not burning too easily for the former and mental strength or high levels of attention for the latter. A more general definition is provided by the World Bank (2021): It is the "ability of a system, community, or society exposed to hazards to resist, absorb, accommodate, adapt to, transform, and recover from the effects of a hazard in a timely and efficient manner, including through the preservation and restoration of its essential basic structures and functions through risk management" (World Bank 2021: 5). As wildfires can spread quickly throughout the land during hot, dry, and windy conditions, communities located near a 'wild', uninhabited area are disproportionately vulnerable. The Canadian Climate Institute or CCI (2020) says that this "wildland-urban interface" (CCI 2020), formally abbreviated WUI and covering approximately 32 hectares of the country, must especially be prepared to face dangerous fires. However, the institute stresses that to do that, support mechanisms in terms of financing and regulation must be put in place – or, if they are already, enhanced.

A particularly devastating and costly fire incident was the one in Fort McMurray in the province of Alberta in 2016. It was an entire "system of wildfires" (Canadian Disaster Database n.d.) that was started by a first one discovered on April 30 (ibid.). Referencing other articles, the CCI describes the multiscale consequences of the fire, reaching from the destruction of homes and displacement of residents to mental health issues and the connection to educational attainment (CCI 2020). The cost of all these individual impacts amounted to some 9 billion dollars back then. Additionally, the fire's strength "destroyed more than 2,400 structures and displaced 85,000 people" (ibid.), making this "the largest evacuation in Canadian history" (ibid.).

After the disaster, fire investigators and other experts conducted analyses of what led to the destruction of so many homes across a substantial geographical space. Through detailed "hazard level ratings" (CCI 2020), they strove to determine what the actual reasons were for this large-scale disaster impact. What they found was that a "simple weakness" could trigger flame infernos, in other words, obvious drivers of fire: "woodpile or shrubs next to the house, gutters with dried leaves, long grass, wood chips in the garden, wood siding, wood decks, or a roof with low fire resistance" (CCI 2020). Obviously, these circumstances make the homes prone to this type of disaster. However, especially in rural areas, this wood is used for heating in cold and rough winter conditions.

Moreover, an important and non-negligible consideration of wildfires is the impact on health, directly conflicting with SDGs #3 (Good Health and Well-Being) and #6, Clean Water and Sanitation (UNDESA n.d.): Whether directly confronted with the fire or farther away, toxic smoke accounts for "reduced lung function, bronchitis, exacerbation of asthma, and increased risk of death" (CCI 2020). Visibly, these issues of what the IAWF calls "symbiotic disturbances" (IAWF 2015) have a temporal component, which drives the necessity of complex, costly, and long-term treatment (IAWF 2015). Additionally, toxic parts of burned materials and other chemicals can pollute people's drinking water. This leads to an increased prevalence of water-borne, gastrointestinal diseases (ibid.).

Apart from wildfires, there is another issue that alarms Canadians: flooding. One might think of them predominantly present in subtropical and tropical regions where these patterns seem to be more common at first glance. However, Canada has large waterbodies and -systems (bays, lakes, rivers and of course the Atlantic and Pacific Oceans), making this a very real problem. In the former three, flooding occurs as soon as the water rises over bed and channel levels, while runoff, storms and waves are the drivers for oceanic flooding (Public Safety Canada n.d.). Moreover, during the long winters in areas such as Alberta, Nunavut and the Northwest Territories, snow, and other kinds of gradually freezing or frozen precipitation melt during the springtime, often paired with heavy rainfall. These risks are decisively exacerbated by climate change patterns.

The heavy rain, however, is not only connected to the warmer season and thawing processes. Thunderstorms and stronger storms can also produce devastating amounts of water, and cause so-called flash floods – in other words, water coming into an area and rising quickly. Mountain ranges, like the Rocky Mountains in western Canada, are a predominant geological driver for flash floods. The same is true for the extensive lakes on the American border, as well as farther north: Often, they are contained naturally (glaciers, soil) or by man-made dams, but water could break free from these limits and engulf into valleys and towns. Further, provinces like Saskatchewan or Manitoba have substantial prairie cover, and snow melts there could also bring about flooding.

2.1. Defeating Heat and Devastation: The FireSmart Program

With these findings, the material for discussion about the reduction of "home vulnerability" (CCI 2020) and that of its inhabitants had been provided and it began or, rather, was re-energized: A non-profit association stretching across and connecting multiple disciplines, Partners in Protection (PiP) from Alberta, developed a program dubbed FireSmart. The primary task of this program is to provide a mechanism for accumulating knowledge and expanding capacity, which is necessary in the affected communities. At the most essential level, a zonal model for home ignition hazard(s) was constructed, motivating homeowners to be more cautious about material they place around houses, yards, and garages. Hence, it serves as a mental reminder for people to decrease the risk of a wildfire threatening their home if they live near an area where it first started.

The way the information is shared is both personally and via refined digital infrastructure. Firefighters conduct professional assessments of individual homes and educate inhabitants of potential risk sources, precautions, and financial arrangements for better protection. For constant improvement purposes, they come back to look at how the situation and environment have changed over time. Digital infrastructure means an app that not only provides possibilities for do-it-yourself (DIY) protective, assessment, and identification measures, but also helps to keep homeowners informed by experts in the field (CCI 2020; FireSmart Canada 2018, 2019). For instance, information on the level of fire resistance is provided in a categorized material overview to facilitate investment and home modification considerations. Surmounting the hurdles of so-called information frictions – unavailable and/or inaccessible information – can be the difference between life and death in a disaster situation like a wildfire.

However, the FireSmart program is not only preparing individual homeowners for wildfires. Through pooling of physical and abstract resources, it enlarges its geographical coverage to entire communities and towns. This is closely tied to SDG #11 (Sustainable Cities and Communities), as disaster prevention has implications for the sustainability and livability of a city. An example of how this plays out is Canmore, Alberta: It has come up with a detailed plan inside FireSmart about how hazardous the conditions are, what materials are prone to ignite, and how you should behave when a wildfire approaches. This has strengthened and detailed the town's overall response mechanism, something which became a model for a national FireSmart program initiative (CCI 2020).

Nevertheless, despite successes that were seen over the years, the FireSmart program implies that both individual people and communities themselves are required to be the instigators for increased safety, something connected to financial issues: The unaffordability of the vast number of new products and technologies is a prominent issue, especially given the demographics in the more rural and northern regions of Canada – lower-income and indigenous populations, for example. For purposes of mitigation and risk reduction efforts, discussions have started about increased governmental involvement, whether from the federal one in Ottawa or the provincial ones. This came after the Canadian Council of Forest Ministers (CCFM) found that most communities do not take an active role in FireSmart (CCFM 2016), presenting a potential problem to the program's salience. Cooperation between different associations tries to address the problem of excessive risk reduction costs for those building homes, as the transition of national regulations to the provincial level would take precious time.

Other issues in which governments play a role are consultation, developmental management, and investment. Special bodies must be in place to adequately, justly, and effectively consult with communities in which wildfire prevention measures are to be implemented. This holds particularly, though not only, for indigenous inhabitants of the land (CCI 2020). Further, urban planning and management practitioners are looking for more opportunities for ever-increasing development of cities. Arguably, the procedural approach commonly referred to as Sustainable Development is often neglected, leaving communities vulnerable to increased ignition risks. Not many governments seem to respect that there are limits to growth and geographical outward expansion, eliminating the line between communities and places in which wildfires occur more often.

Lastly, investments have largely been funneled to fund response and recovery, and not high enough (IAWF 2015; CCI 2020). However, to be in line with Sustainable Development and the human right of 'freedom from harm', money must also be made available for paying pre-disaster measures to decrease their probability. This would not only be in line with SDG #16 – Peace, Justice and Strong Institutions (UNDESA n.d.) – but would also guarantee people's safety and a healthy life. In other words, as the IAWF put it in 2015, all observable variables of wildfires must promote a "fiscal logic for funding cost-effective mitigation activities" (IAWF 2015). Cost effectiveness, the association says, does not mean to consistently increase the number of areas where mitigation measures are put in place: Instead, it means to look more after areas which have already been the target of investment and prevention. Disproportionate financial resource allocation and overall preventative treatment must be dealt with in a way to limit instead of increase costs of management and rehabilitation. This means to "remove legislative, bureaucratic and market impediments" (IAWF 2015).

In sum, while extensive programs like FireSmart do provide good tools and resources for wildfire risk reduction, the governmental, community-level, urban and social institutions must be further strengthened to correctly assess threats to the population and respond to them in time. Without robust financial, social, and political institutions, the risk of wildfires will remain a consistent threat for many towns and communities on the brink of the rural landscape.

2.2. Flood Protection Mechanisms: Red River Flooding

The Red River Flood occurred in 1997 and was "the largest flood event on the Red River in Manitoba in 145 years" (Burn 1999), covering 141,900 hectares in total (Statistics Canada n.d.). While no one was killed, the flood did not only cause material/physical damage amounting to over $500 million, but also impacted the cohesion of society (ibid.). As alluded to, the triggers for such tremendous flooding events are irregular and high precipitation levels, so occurred during fall 1996 (Burn 1999: 3452) and then during the "unusually long and unusually cold" winter of 1997 (Burn 1999: 3453). These conditions drove a heightening water content in the overall snow layers, and a blizzard caused further rise in water capacity, as it hit once the first melting was observable (Burn 1999: 3453).

Already early on, the reliance on institutional support and effective policy planning was observable, as the Province of Manitoba asked the country's Armed Forces for help on April 19, prompting 'Operation Assistance' (International Red River Basin Task Force 1997: 16). The severely affected City of Winnipeg, moreover, has put a comprehensive evacuation plan in place on April 30 (Burn 1999: 3453). Furthermore, an important institution was the international news media, broadcasting necessary information as well as early warnings about the geographical variability of flood impacts. A "senior water resources engineer with the Manitoba Water Resources Branch" (Burn 1999: 3454) was tasked with this important duty. Given that technology was limited in scope at the time, the accuracy of forecasts is deemed relatively high (Burn 1999: 3454). This activity is what makes Burn (1999) stress some essential components of how people will individually perceive risk. In the media context, the most important ones are organization and selection of received information, as well as the variability of impacts according to specific personal and life characteristics (Burn 1999: 3451).

Flooding is not a new phenomenon in a large part of the area, hence the towns and cities located here have flood protection infrastructure in place in the form of dikes and floodways (International Red River Basin Task Force 1997: 35). This not only bars water from overflowing certain areas, but also diverts large quantities to the rural places to re-flow into original rivers (Burn 1999: 3453). This shows that infrastructural planning had focused on flood prevention for a longer time, emphasizing that community protection was high up on the agenda of the responsible institutions. Thus, they acted in a coherent and attentive way over the years, rather than neglecting necessary steps to prevent disaster.

However, as Burn (1999) reveals, the institutional framework in the Canadian portion of the Red River Valley was subject to harsh criticism. The April 19 forecast was accurate, but too late for some residents to be able to adequately respond to it, particularly in regions other than the southern part of the province without experience (cf. Burn 1999: 3456). In other words, the time to prepare was limited. Further, the manner of ordering the evacuation by the Province of Manitoba was far from widely accepted, and hence many people did not take the requested steps and remained in their houses (Burn 1999: 3455). Obviously, this caused alarm inside emergency management circles. It is a pattern also seen with previous natural disasters, such as the Ahr River Valley flooding in northwestern Germany or the various hurricanes in the southern United States.

Thus, this underlines that the institutions responsible for civil protection and disaster prevention must be working in cooperation with the public to be effective and considerate. Otherwise, this is going to undermine the entire purpose of organizational capacity and response. However, it must be noted here that the criticism was directed to persons and agencies at the provincial level, not at the local and community levels. These were reacting accordingly as they had prior experience in adequate response.

Because of this regionalized variability, there is a necessity for high institutional responsiveness levels, just because areas that have not dealt with prior extreme floodings suffer from limited information and material resources. There is also a demand for further sensitization of the extreme disconnection of flood patterns from one to the next: If one flood was not too severe and preparation time was adequate, this does not have to hold for an upcoming one (see Burn 1999: 3456). This circumstance calls for the more timely and comprehensive collection of flood-relevant data, such as precipitation levels, wind/storm velocity, soil moisture content and more variables.

3. Country: Bangladesh

The countries of the Indian Subcontinent are, as well as significant parts of southeastern Asia, a region of high risk for severe storms. The powerful winds strike the very areas which do not have the necessary infrastructure in place to be protected, and cities and communities most often are positioned at low level. As became evident through storms elsewhere in the world, it is not only the wind that is destructive, but particularly the storm surges that follow.

Bangladesh is prone to both severe storms and their negative effects. Luxbacher and Kamal Uddin (n.d.) say that around 10 million citizens experience one or even more disasters on an annual basis (Luxbacher and Kamal Uddin n.d.: 1). The "deltaic country" (ibid.) is located at lower levels (ibid.: 2), and since the Ganges-Brahmaputra-Meghna (GBM) River Basin has substantial quantities of water that could be absorbed in hot and humid conditions, very severe tropical storms – known as cyclones – are an existential threat to people living here (see International Rivers n.d.). In fact, already in 2004, the United Nations ranked Bangladesh as the one nation mostly at risk for this disaster type (Luxbacher and Kamal Uddin n.d.: 1). One of the deadliest cyclones was the Bhola cyclone in 1970, but also Sidr in 2007 and Aila in 2009 were devastating (International Federation of Red Cross and Red Cresecent Societies 2018: 58). With this danger being prevalent, the Bangladeshi government had to react and implement preparedness strategies.

The coming section will look at a programmatic policy effort that has been launched some decades back, complemented by those initiated after the two cyclones in the mid-2000s. Approaches the government took were changed because of additional funding and resources that became available for the different ministries in Dhaka. Over time, a conglomerate of institutions with different tasks has been set up, all to improve not only response and recovery mechanisms after storms, but also to use technologies and materials to better prepare for them. The program will be explained and assessed, broadening the understanding of how the thoughts about disaster risk reduction have shifted in the country.

3.1. Cyclones and their Management

The Bangladeshi government has established major cooperative programs over the last decade, two of which deserve consideration: One that was started in 1972 joining hands with the Bangladeshi Red Crescent Society (BDRCS) and another prioritizing the strategic development of disaster risk reduction and preparedness policies.

To adequately describe the concept of preparedness, there are two definitions provided in a report by the World Bank (2021). In United Nations Disaster Risk Reduction (UNDRR) parlance, preparedness would be the "knowledge and capacities developed by governments, response and recovery organizations, communities, and individuals to effectively anticipate, respond to, and recover from the impacts of likely, imminent, or current disasters" (World Bank 2021: 5). A different, more active phrasing is used by the Directorate-General for European Civil Protection and Humanitarian Aid Operations, for short DG-ECHO: "Organizational activities which ensure that the systems, procedures, and resources required to confront a natural disaster are available to provide timely assistance to those affected, using existing mechanisms wherever possible" (World Bank 2021: 6).

Hence, the catchphrase of preparedness became a steering force for major investments by the government, which made the building of cyclone shelters and the establishment of effective early warning systems (EWS) possible (Luxbacher and Kamal Uddin n.d.: 2). Not only was the realization of both susceptibility and severity important for disaster management decisions, but also past cyclones in the late 1990s. Earlier initiatives were not successful, and because of what happened during these cyclones, the corresponding steps could be taken to build capacity and to foster more infrastructural, scientific, and human capital.

The first phase of the so-called Comprehensive Disaster Management Programme, or CDMP, was endorsed in 2003 with funding from the United Nations Development Program (UNDP), the British Department for International Development (DFID) and the United Nations Office for Project Services, UNOPS (ibid.: 3). The authors note that over the years of the program, several boards, committees, bureaus, and task forces in the realm of disaster management have further consolidated and intensified monitoring, advisory, cooperation, and planning efforts (ibid.: 4). In the institutional summary, it becomes apparent that the Bangladeshi government tried to unify several actors in the sphere, from officials to non-governmental organizations (NGOs) to the community level and mayoral representation in decisions.

The whole phase I ran from 2004 to 2009, involving several workshops and presentations to sensitize people, to educate them about the dangers that cyclones bring. This initiation phase was intended to set the precedent for policy innovation – i.e., to firmly incorporate disaster management approaches not only into people's minds and lives, but into political rhetoric and decisions. As mentioned, previous experience prompted more action(s), which explains a substantial geographical extension of the program coverage (ibid.: 6).

However, implementation was delayed several times, whether through other climatological catastrophes like flooding, because of political unrest or lack of effective coordination in the governmental branches (ibid.: 7). The last point was both because of personnel volatilities in the responsible ministries and refusal of certain officials to complete the amount of work necessary for adequate and long-term implementation. Moreover, there were indeed skeptics present who voiced their discomfort with the program. Luxbacher and Kamal Uddin do not specify those, but they underline that it was time-consuming to get some population groups convince that the program would eventually be beneficial (ibid.: 7).

As the authors elaborate, CDMP Phase I was able to create a multitude of funding, regulatory, awareness and policy mechanisms that surely transformed the governmental and community approach of disaster risk reduction, preparedness, and response in Bangladesh. The

Climate Change Cell in the Department of the Environment inside the Ministry of Environment and Forests, for instance, makes sure to collect and publish necessary data relevant for Bangladesh as a whole.

Monitoring and reporting of emergency situations is taken over by the Disaster Management Information Center, providing a day-round service when an emergency occurs (ibid.: 7). Through information and communication technology, the government as well as cities and communities (and therefore, the people) could communicate with each other to be adequately informed of all actions and necessities. This was mostly achieved through Community Risk Assessments (CRAs), an extensive process of evaluating the situational needs of the population (ibid.: 8). On pages eight to twelve, the authors further outline government-academic cooperation with the Bangladesh University of Engineering and Technology, as well as provide information on the outcomes of phase II of the program.

The country's Red Crescent Society was an instrumental actor for the Cyclone Preparedness Programme (CPP) that is cooperatively run with governmental branches since 1972 (International Federation of Red Cross and Red Crescent Societies 2018: 57) but proved particularly vital in 2017 when Cyclone Mora came. The program was intended as an initiative cognizant of the country's communities, especially those along the Bay of Bengal shorelines. The essence of the program were volunteer workers who walked, biked, or rode their motorcycles through communities when it was determined that a natural disaster was imminent. For increased visibility, and thus more reach, they used colored signaling flags, a symbolic practice quite common in early warning activities. However, radio transmissions were another crucial means of spreading the word as well, especially because the overall systems are not as efficient as in other countries. Obviously, winning volunteers for this kind of important work requires profound training, which is routinely provided by BDRCS.

The aim of this program, as with all disaster management programs, the actions by volunteers were directed toward effective and large-scale vulnerability reduction of the affected population (International Federation of Red Cross and Red Crescent Societies 2018: 58). This was done very effectively in the wake of cyclone Mora at the end of May 2017, as warning efforts stretched "across a population area covering 11 million people" (ibid.), whereas 500,000 could be taken to safe locations in less than a full day. What played a huge part here was the ability of the volunteers along with so-called Community Disaster Response Teams (CDRTs) to bring the messages across in the local language. This helped to provide all the necessary information and precautions people needed to take (see ibid.).

The program had measurable impacts, as it substantially reduce deaths caused by cyclones over time: While a cyclone in 1970 cost half a million lives, those in 2007 and 2009 showed numbers of 4,000 and 190 (International Federation of Red Cross and Red Crescent Societies 2018: 58). This was possible because of an "[a]ll-of-society-engagement" (ibid.), which provided a holistic approach to disaster management, preparedness, and recovery.

However, the program also faced challenges, which were both material and social: Signal flags have no lights, which renders them invisible at night and they are neither immune to high winds. The social side is general safety concerns in shelters, particularly voiced by women and girls (International Federation of Red Cross and Red Crescent Societies 2018: 59). Nevertheless, the cooperative program has proven effective, also because it reduced negative spillovers, being damage of property and livestock herds (ibid.). Hence, we can say that through rigorously following SDG #17 (Partnerships For The Goals) the country has become substantially more adapted to the threat and effects of cyclones.

4. Country: Indonesia – Climate Change in Southeast Asia

Indonesia is made up of several islands stretching along a region known as the 'Ring of Fire.' This term describes a geographical space where both tectonic and volcanic activity can be observed on a more regular basis. Because of these two factors, Indonesia has experienced a few earthquakes in the past. They have a considerable impact on communities and cities, as the infrastructure is destroyed by the shaking earth. During these events, several people lost their lives due to falling trees, debris, or shocks by torn power lines.

Evidently, the threat of earthquakes is by far not the only one, also because the country likewise experiences the negative externalities of worsening climate change. A major threat, as seen over the last few months, is torrential rainfall which causes both flooding and dangerous landslides in which houses are destroyed and people encapsulated. This reality, as in Bangladesh, is further exacerbated by the propensity for strong cyclones or typhoons hitting the area. Winds can pick up speed and destructive force quickly, posing a risk to both human lives and infrastructure. This holds particularly for rural places, of which there are many in the country.

4.1. Earthquakes – A Continuous Tectonic Uncertainty

A particularly devastating earthquake event occurred in 2018 in the East Lombok province. Its impact prompted the government in Jakarta to take more pronounced action toward earthquake preparedness: The Civil Protection Agency of Indonesia has been pooling resources over the last few years to improve the resilience and sustainability of communities and towns across the country, which has already shown effects. This occurs in the context of SDG #11, Sustainable Cities and Communities. Education on how to behave during earthquakes, for example, is a key part in sensitizing the population for this issue, and an essential of SDG #4 (Quality Education). Of course, raising awareness alone will not help Indonesians to respond to earthquakes.

A particularly important process examined in the case study by Kusumastuti et al. (2021) is knowledge management in the context of disaster risk reduction and relief activities. Of course, this implies that the necessary platforms for the acquisition of knowledge must be in place and functional enough to share it effectively. In other words, people rely on reliable institutions, something that is codified in SDG #16 (Peace, Justice and Strong Institutions). By interviewing representatives of non-governmental, governmental, and international organizations, authors would first inquire about Indonesian disaster management practices. Later, they would compile a questionnaire to ask people who experienced the earthquakes *in-situ* how they used the gained knowledge for their own actions (Kusumastuti et al. 2021: 4; see also Pribadi et al. 2021).

To underline the importance of knowledge in this situation, authors explain the different spheres of knowledge sharing, that in can occur one-way or both ways. In other words, it is either transmitted to the people from 'the outside' through media, or it is shared by external actors but in close cooperation with the outside information providers. Also, knowledge can be transmitted in the community and thus given the ability to circulate within it (see ibid., pp. 5-7). This was an important step in the process of dealing with the earthquakes that hit Lombok.

The study shows that both when the earthquake began and when it stopped, a high number of people showed correct behavior according to what they have been taught under the initiated disaster management programs (ibid.: 9). Hence, they were able to disclose the efficacy and efficiency of the programs, and thus the actions were following SDGs #4 (Quality Education), #11 (Sustainable Cities and Communities; with targets 11.5 and 11b), #13 (Climate Action; with targets 13.1-3) and of course #17, Partnerships For The Goals (The Danish Institute for Human Rights n.d.).

Of course, technological and organizational challenges remain to this day. However, if planners and government officials can work together and extend this to the respective communities as well, all actors can build this knowledge which then can be transferred to walk a multimodal path. To be prepared for future sudden-onset disasters, Indonesia might also do good to appeal to foreign countries for assistance in mitigation planning, and particularly speak to fellow members of the Association of Southeast Asian Nations (ASEAN) to receive incentives and ideas for further resource allocation, planning and adequate decision-making.

5. Conclusion

Disasters can strike in many given forms, and they are decisively exacerbated in scope, scale, and intensity by the negative externalities of climate change patterns. This paper has given an overview about three countries and the various mechanisms they use to prepare for, address, and overcome the devastation of wildfires and floods (Canada), cyclones (Bangladesh), and earthquakes (Indonesia).

The description of the different countries made clear that the whole planet faces natural disaster, not matter what climatological situation prevails in a place. Evidently, some places are more prone than others to certain types of disaster, but climate change has shifted the geographical distribution of the likelihood of disasters – and with that the vulnerability patterns of cities, communities, regions, and entire countries.

Because of this reality, countries have activated their institutional capacities, or have built new ones over time. These were necessary because of limited experience, funding constraints, resource availability and available human capital. The observations in the management of the disasters are consistent with the argument made in this paper: Responsive institutions working across sectoral and other structural or cultural boundaries are necessary – indeed, indispensable – for countries to protect their citizens from the harm of natural disasters in all forms. They must be capable of adaptation, amelioration, and innovation, which implies a constant monitoring and evolution of organizational structures. Only if dynamic and resilient institutions exist in a country, the process of disaster management and risk reduction can be a successful one.

References:

British Columbia Centre for Disease Control (2020). "Wildfire Smoke and Your Health."
BCCDC, online at: http://www.bccdc.ca/resource-
gallery/Documents/Guidelines%20and%20Forms/Guidelines%20and%20Manuals/He
alth-Environment/BCCDC_WildFire_FactSheet_HowToPrepare.pdf.

Burn, D.H. (1999). "Perceptions of flood risk: A case study of the Red River flood of 1997."
Water Resources Research 35 (1), available at:
https://agupubs.onlinelibrary.wiley.com/doi/pdfdirect/10.1029/1999WR900215.

CCFM (2016). "Canadian Wildland Fire Strategy: A 10-year Review and Renewed Call to
Action." Canadian Council of Forest Ministers.
https://cfs.nrcan.gc.ca/publications/download-pdf/37108.

Calkin, D.E., Cohen, J.D., Finney, M.A. & Thompson, M.P. (2013). "How risk management
can prevent future wildfire disasters in the wildland-urban interface." Retrieved from:
https://www.pnas.org/doi/full/10.1073/pnas.1315088111.

Canadian Climate Institute (2020). "Enhancing Community Resilience as Wildfire Risk
Increases." CCI, online at: https://climateinstitute.ca/publications/enhancing-
community-resilience-as-wildfire-risk-increases/.

Canadian Interagency Forest Fire Centre (2021). "Canada Report 2021." CIFFC, online at:
https://www.ciffc.ca/publications/canada-reports.

Canadian Disaster Database (n.d.). "Assiniboine, Red and Winnipeg Rivers MB."
Government of Canada/Gouvernement du Canada, Public Safety Canada. Online at:
https://cdd.publicsafety.gc.ca/dtpg-eng.aspx?cultureCode=en-
Ca&boundingBox=&provinces=3&eventTypes=%27FL%27&eventStartDate=&injure
d=&evacuated=&totalCost=&dead=&normalizedCostYear=1&eventId=266.

Canmore (2020). "FireSmart." Town of Canmore, online at: https://canmore.ca/municipal-
services/emergency-services/emergency-management/firesmart.

Canada Wildfire (2018). "Mapping Canadian wildland fire interface areas." Retrieved from:
https://www.canadawildfire.org/mapping-wui.

The Danish Institute for Human Rights (n.d.). "Goals, targets and indicators." The Human
Rights Guide to the Sustainable Development Goals, online at:
https://sdg.humanrights.dk/en/goals-and-targets.

Global Facility for Disaster Reduction and Recovery (n.d.). "Indonesia - investing in urban
flood resilience and seismic risk mitigation." GFDRR, online at:
https://www.gfdrr.org/en/indonesia-investing-urban-flood-resilience-and-seismic-risk-
mitigation.

Heinrich Böll Foundation Southeast Asia (2022). "Disaster Management in Indonesia:
Complex Challenges of a Dual Early Warning System." Retrieved from:
https://th.boell.org/en/2022/03/23/disaster-management-indonesia.

International Association of Wildland Fire (2015). "Reduce wildfire risks or pay more for fire disasters." IAWF, online at: https://www.iawfonline.org/article/reduce-wildfire-risks-or-pay-more-for-fire-disasters/.

International Federation of Red Cross and Red Crescent Societies (2018). "CPP Early Warning: Saving Thousands in Cyclone Mora." Case Studies: Red Cross Red Crescent Disaster Risk Reduction in Action – What Works at Local Level. IFRC, available at: https://reliefweb.int/report/world/case-studies-red-cross-red-crescent-disaster-risk-reduction-action-what-works-local.

International Red River Basin Task Force (1997). "Red River Flooding Short-Term Measures." Interim Report of the International Red River Basin Task Force to the International Joint Commission, December 1997, Ottawa/Washington. Available at: https://www.ijc.org/sites/default/files/2018-10/RR%20Flooding%20report.pdf.

International Rivers (n.d.). "Ganges-Brahmaputra-Meghna." Online at: https://www.internationalrivers.org/where-we-work/asia/ganges-brahmaputra-meghna/.

Pribadi, K.S., Abduh, M., Wirahadikusumah, R.D., Hanifa, N.R., Irsyam, M., Kusumaningrum, P. & Puri, E (2021). "Learning from past earthquake disasters: The need for knowledge management system to enhance infrastructure resilience in Indonesia." *International Journal of Disaster Risk Reduction*. DOI: https://doi.org/10.1016/j.ijdrr.2021.102424.

Kusumastuti, R.D., Arviansyah, A., Nurmala, N. & Wibowo, Sigit S. (2021). "Knowledge management and natural disaster preparedness: A systematic literature review and a case study of East Lombok, Indonesia." *International Journal of Disaster Risk Reduction*. DOI: https://doi.org/10.1016/j.ijdrr.2021.102223.

Luxbacher, K. & Kamal Uddin, A.M. (n.d.). "World Resources Report Case Study. Bangladesh's Comprehensive Approach to Disaster Management." World Resources Report, Washington DC. Available online at: http://www.worldresourcesreport.org.

Natural Resources Canada. "Forest Fires." Government of Canada/Gouvernement du Canada, online at: https://www.nrcan.gc.ca/our-natural-resources/forests/wildland-fires-insects-disturbances/forest-fires/13143.

Natural Resources Canada (n.d.). "Climate Change and fire." Government of Canada/Gouvernement du Canada, online at: https://www.nrcan.gc.ca/our-natural-resources/forests/wildland-fires-insects-disturbances/climate-change-fire/13155.

Province of Manitoba (n.d.). "Historic Flood – 1997." Retrieved from: https://www.gov.mb.ca/mit/wms/rrf/historical_1997.html.

Statistics Canada (n.d.). "Red River flooding." Available online at: https://www150.statcan.gc.ca/n1/pub/11-402-x/2011000/chap/geo/geo04-eng.htm.

The World Bank (2021). "How Indonesia Strengthened its Disaster Response with Risk Finance and Insurance." Retrieved from:

https://www.worldbank.org/en/news/feature/2021/11/17/how-indonesia-strengthened-its-disaster-response-with-risk-finance-and-insurance.

The World Bank (2021). "Understanding the Needs of Civil Protection Agencies and Opportunities for Scaling up Disaster Risk Management Investments." World Bank Group, available at: https://openknowledge.worldbank.org/handle/10986/36292.

United Nations Department of Economic and Social Affairs (n.d.). "Do you know all 17 Goals?" UNDESA, online at: https://sdgs.un.org/goals.

Westhaver, A. (2017). "Why Some Homes Survived: Learning from the Fort McMurray wildland/urban interface fire disaster." Retrieved from: https://www.iclr.org/wp-content/uploads/PDFS/why-some-homes-survived-learning-from-the-fort-mcmurray-wildland-urban-interface-fire-disaster.pdf.

www.ingramcontent.com/pod-product-compliance
Lightning Source LLC
LaVergne TN
LVHW041818190726

843493LV00009B/2949